Michael Herlache MBA
Doctor of Business Administration Candidate
AltQuest Group & Founders Ventures

Billionaire Handbook

Combining Wall Street & Silicon Valley to Build Billion Dollar Companies

For those making the transition from the sell-side and the buy-side to the build-side.

About the Author:

Michael Herlache is a VP of M&A at AltQuest Group based out of Fort Lauderdale/Miami, FL. He is also the CEO of Founders Ventures, a startup lab, pre-accelerator fund, and accelerator. He lives in his home in Florida with his beautiful wife. Michael has an MBA in Finance and is getting his Doctorate in Business Administration.

He is on the Board of Directors of M&A Nexus, Asiansbook, FundLinked, DegreeLinked, NationLinked, and FameLinked.

To learn more about Founders Ventures, please go to www.VCFounders.com:

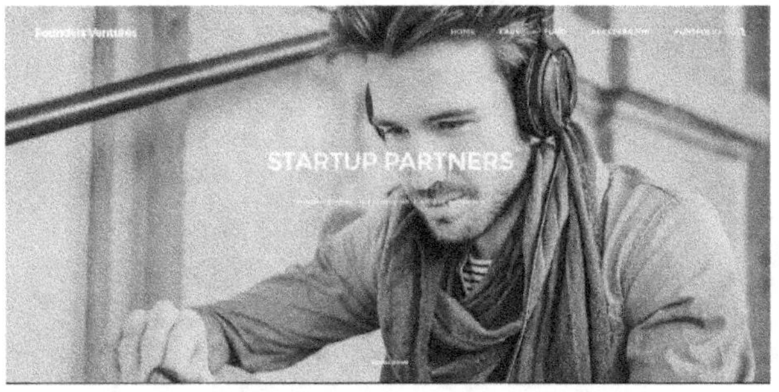

To learn more about AltQuest Group, please go to www.AltQuest.com.

Contents

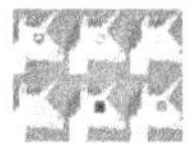

Preface

Buy-side. Sell-side. Build-Side?

Buy-side (buying perpetuities)
Sell-side (selling perpetuities)
Build-side (building perpetuities) - combining Wall Street valuation with Silicon Valley platform building

We have all heard about the buy side and the sell side of finance, but who is actually building the perpetuities. What if there was a third side of finance? What if there was a build-side, with individuals possessing IB/PE and platform development talents to use in building perpetuities? Shouldn't that be the logical course of events with individuals taking their knowledge of valuation and industries and putting them to use in building the next unicorns?

So what would this look like? IB/PE professionals joining startup labs such as the one I run called Founders Ventures (www.VCFounders.com) to work

on concepts that have a legitimate chance of being a unicorn. Rather than leaving one's job to join a questionable startup, join a startup lab and be directly involved in the build-side, even if it part-time. The work of the build-side is syndication.

Shouldn't we all be working towards getting on the build-side?

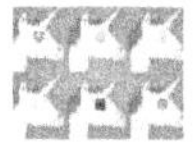

Chapter 1: Introduction to the Unicorn Methodology

Unicorn Methodology

Concept:
Market size:
Total addressable market:
UNIT ECONOMICS (PROFITABILITY) OF THE COMPANY:
a. IDENTIFY COST OF ACQUISITION PER CUSTOMER (CAC):
b. IDENTIFY COST PER USER PER MONTH:
c. IDENTIFY REVENUE PER USER PER MONTH:
d. IDENTIFY LIFETIME VALUE PER USER (LTV):

MODEL PLATFORM ARCHITECTURE IN TERMS OF BASE PHP PLATFORM AND MODS ONTO IT. DOMAIN WITH MVP DONE: www.domainnamehere.com
DATABASE OF 25 HIGHEST PRIORITY USERS OF PLATFORM AND COLLECT EMAIL, NAME, NUMBER. .XLS NAME WITH TOP 25 DATABASE: databasenamehere.xls
SELF-ADVOCATE TO THE 25 HIGHEST PRIORITY USERS OF PLATFORM AND NOTIFY THEM THAT YOU WILL BE SENDING THEM VALUE AND THAT THIS IS THEIR ACCOUNT TO CLAIM IT GIVING THEM LOGIN CREDENTIALS. .DOC NAME WITH MESSAGE TO BE SENT: companyname.doc
SCREEN SHARE WALK THROUGH WITH THE HIGH PRIORITY USER AND SHOW THEM HOW IT WORKS AND GET THEM TO LOGIN BY THEMSELVES
USER UTILIZES PLATFORM BY THEMSELVES FOR A TRIAL PERIOD OF TIME (3 TO 6 MONTHS)
SEND USER THE INVOICE FOR USING THE PLATFORM. FEE SHOULD BE MEMBERSHIP FEE BASED UPON STANDARD MARKET FEE. AMOUNT OF FEE & INVOICING METHOD: $XXX.00 per month & PayPal
BUILD DATABASE OF 26-50th HIGHEST PRIORITY USERS OF PLATFORM AND COLLECT EMAIL, NAME,

NUMBER. .XLS NAME WITH TOP 26-50 DATABASE: databasenamehere2.xls
IDENTIFY 3 TO 5 MOST LIKELY STRATEGIC BUYERS OF YOUR PLATFORM. REACH OUT TO THE PERSONALLY WHEN YOU HAVE $100M IN RECURRING REVENUE AND EXIT AT THE MARKET REVENUE MULTIPLE BASED UPON PRECEDENT TRANSACTIONS: Facebook, Google, LinkedIn, Alibaba

DISCOUNT RATE STATUS:
Seed: pre-database MVP: 80% discount rate
Angel: self-advocacy & activation: 50%-70% discount rate
Series A: utilization & monetization: 40%-60% discount rate
Series B: monetization & scale: 30%-50% discount rate
Growth: Scale & new markets: 20%-40% discount rate
Mezzanine: Scale in old & new markets: 15%-30% discount rate
IPO: Scale globally: 7-15% discount rate
Post-IPO Minority Ownership: Mature & New Product Development: 3.5%-10% discount rate

FINANCIAL MODEL & VALUATION (DCF):

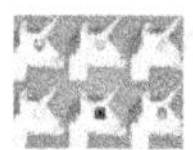

1) We found 84 U.S.-based companies belong to what we call the "unicorn club," a jaw-dropping 115% increase from our last post. The increase is driven largely by "paper unicorns" – private companies that have not yet had a "liquidity event." But, these companies are still a super-rarity: our list is just .14% of venture-backed consumer and enterprise tech startups.

2) On average, eight unicorns were born per year in the past decade (versus four in the 2003-2013 era). There's not yet a super-unicorn ($100 billion-plus in value) born from the 2005-2015 decade, but there are now nine "decacorns" ($10 billion-plus in value), 3x our last post.

3) Consumer-oriented companies drive the majority of value in our set: more companies and higher average value per company. They raise a lot of private capital.

4) Enterprise-oriented companies are fewer and raise less private capital; and increased enterprise

fundraising has reduced their return on private dollars raised.

5) In terms of business models, e-commerce companies drive the majority of value in our set, but have the lowest "capital efficiency." Enterprise and audience companies have decreased in market share of our set, while SaaS companies have grown in market share significantly. We've also added a new category: Internet of Things/consumer electronics.

6) It's a long journey, beyond vesting periods: it has taken ~7 years on average before a "liquidity event" for the 39% who have 'exited' – not including the 61% of our list that is still private. The capital efficiency of these "private unicorns" is surprisingly low, which will likely impact future returns for founders, investors and employees.

7) Take heart, "old people" of Silicon Valley: Companies with educated, tech-savvy, experienced 30-something, co-founding teams with history together have built the most successes. Twenty-something founders and successful pivots are the

minority; dedicated CEOs who are able to scale their companies for the long haul are not.

8) San Francisco maintains dominance as the new epicenter of the Bay Area's most valuable tech companies; cities like NYC and L.A. are growing in importance

9) Immigrants play a huge role in the founding and value creation of today's tech companies. We wonder how much more value could be created if it were easier to get a work visa.
We identified 84 companies for our set (by our definition, U.S.-based, VC-backed software and Internet-oriented companies founded since 2005 and valued at over $1 billion by public or private market investors1). That's a staggering 115% increase since our last analysis just a year-and-a-half ago.
The total value of these companies is $327 Billion – 2.4x our last analysis (excluding Facebook, which was almost half the value of our last list).
It's the number of companies, not their individual valuations, driving the dramatic increase in total value. The average company value on our list is worth $3.9 billion, just an ~8% increase from last time.

And it's the number of "paper unicorns" that has dramatically increased the total value. Private companies are now 61% (vs 36%) of our list, worth $188 billion in total and $3.7 billion on average.
Why so many more 'unicorn' companies now versus 2013? Some thoughts:
a) Compelling products that are easier than ever to adopt through large and growing global markets, smartphones, and social networks to spread the word faster. This is driving more exciting growth, adoption and engagement numbers than ever before (A16Z's strong presentation on the underlying fundamentals here.

b) A perception of winner-take-all markets due to branding, scale and/or network effects, and intimidating, growing cash war chests (see here) driving investor FOMO, and demand to invest in 'winners' at almost any price (Bill Gurley has great insights about that here).

c) Competitive later stage capital from more sources than ever – late stage funds, public investors investing earlier, and global strategics. These investors often have a low cost of capital which gives them a lower

return hurdle than traditional venture investors; and they often receive downside protection as part of their investment that isn't reflected in valuations (Fenwick's great analysis of this trend here).

d) Vibrant public markets fueling optimism: the NASDAQ is up 32% since our last analysis.

e) Optimistic private markets sheltering a thicket of "paper unicorns." When companies are private, founders can share more about their future dreams with investors; report less; and the shares are illiquid, constraining short-term changes in valuation. Those factors, combined with the above reasons, have driven significant 'multiple inflation' for private versus public company valuations (giving private companies more value than public ones as a multiple of # of engaged users, revenues, EBIT, growth rate), including from the WSJ;
The Verge; Josh Kopelman; Thomas Tunguz; CB Insights; and Deepak Ravchanran.

Despite the doubling, building one of these companies is still ridiculously difficult and rare. If 60,000 software and Internet companies were funded

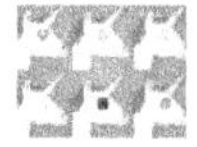

in the past decade2, that means only .14% have become unicorns– or one in every 714. The odds of building, working for or backing one are worse than catching a ball at a major league game; but, better than the chance of dying by shark attack – so we've got that going for us, which is nice.

2) On average, eight unicorns were born per year (versus four in the 2003-2013 era) in the past decade. There's not yet a super-unicorn ($100Bn+ in value) born from 2005-2015, but there are now nine "decacorns" ($10 billion-plus in value), 3x our last post.

unicorn-companies-year

The best years to start a unicorn were 2007 (27% of companies in our set) and 2009 (18%). During these 2 years, 45% of the companies in our set were started. What happened in 2007 and 2009? 2007: The launch of the iPhone. By the end of the year, the beginning of the worst US economic crisis since the great depression; 2009: Just months previously, Android was launched; and the bottom of the "great recession" occurred – the lowest point of the NASDAQ, S&P500 and DJIA in the last decade

The best times to start a unicorn company could be a) post the launch of a watershed new tech platform; and b) during a prolonged public market downturn. Without many great jobs available, the reduced opportunity cost and related hardship may spawn great innovation and grit.

To note, there are some astoundingly young companies in our set including Illumio, Oscar Health and Zenefits, all <3 years young. The combo of great products plus growing global, mobile markets may have accelerated time-to-'escape velocity' adoption and correspondingly, time-to-unicorn for some companies. This rapid ramp could also mean a rapid decline for other companies if customers are not super engaged and happy for the long haul.

As we wrote previously, every major technology wave has given birth to one or more "super-unicorns" – companies that grow to be worth >$100B over time. Facebook, the super-unicorn of the 2000s, has dramatically increased in value – it is now worth $247Bn, up 102% since we last wrote. Now they are worth more than the sum of all other companies on last year's list, and all the consumer companies on our current list. (PS: FB has also 'aged out' of our analysis, as it's now 11 years old).

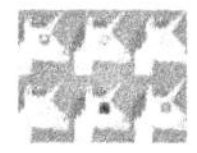

There are now 9 "decacorns" (in our lingo, companies worth >$10bn), a 3x increase vs. our last analysis. To note, 5 of the 9 are largely mobile (Uber, Twitter, WhatsApp, SnapChat, Pinterest).

History suggests the 2010s will give rise to a super-unicorn or two that reflect the key tech wave of the decade, the mobile web. Whichever company (or companies) comes to represent this key innovation (Uber?) will likely continue to accelerate in value as FB, Google and Amzn have over the past decade.

3) Consumer-oriented unicorns continue to drive the majority of value in our set: more companies, and higher average value per company. They also raise a lot of private capital.

Consumer-oriented companies (companies where the primary customer is a consumer) contribute 72% of the aggregate value on our list (vs 60% last time), and comprise 55% of the companies on our list. They are worth $5.1Bn on average.

8 of the top 10 most valuable companies are consumer–oriented; consumer companies seem to reach higher peaks than enterprise companies.

They are currently worth about 11x the private capital raised on average (excluding >100x outliers

WhatsApp, FitBit and YouTube; also 11x on average in our last analysis. 9x is the current median)
When founders start a company aiming to be super successful, they may not realize how many rounds of dilution may be ahead; or how many dollars of liquidation preference might be added (Heidi Roizen wrote an excellent post on this here).
Cases in point – the consumer companies on our list have raised on average $535 million in six private rounds (that's series E or beyond) versus $348 million in our last analysis, a whopping 54% increase. Seven consumer companies on our list have raised over $1 billion in private capital each.
Some amazing exits have been driven from consumer-oriented companies relative to private capital raised (and more than half through acquisition): WhatsApp (325x!), YouTube (144x), Fitbit (132x), Zillow (48x), and Nest (40x)
Conversely, 20% of our consumer-oriented companies are valued at <4x the private capital raised: Evernote, FanDuel, Gilt Groupe, Groupon, JustFab, Lyft, SoFi, Tango, and Zynga
4) Enterprise-oriented companies raise much less private capital; but increased fundraising has seriously reduced their capital efficiency.

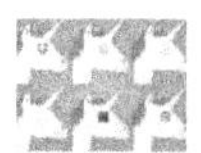

The average enterprise-oriented company (where the primary customer is a business) is worth $2.5 billion, less than half the average consumer company.
To note, their private capital raised is $247 million on average, up 79% versus our last analysis. This increase has reduced their average capital efficiency significantly; from 26x in our last analysis, to 7.6x for the median enterprise company in our set.
Some outstanding enterprise-oriented exits relative to private capital raised: Veeva, Workday, and Softlayer (926x!, 87x, 67x respectively)
Conversely, 18% of our enterprise-oriented companies' recent valuation is <4x their private capital raised: AppNexus, Automattic, Box, Cloudera, Lookout, MagicLeap and Simplivity.
5) E-commerce drives the most value of five primary business models, and SaaS companies have significantly increased their market share since our last analysis. We've also added a fifth business model category: Consumer Electronics/Internet of Things.

E-Commerce companies (companies where a consumer pays for a good or service through the

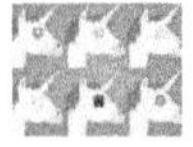

internet or mobile; including companies like Uber and Airbnb) continue to drive the most value (36%); they also raise the most private money ($683m on average!), and deliver the lowest multiples (8x average) of valuation over capital raised. This is likely because of increased headcount and marketing costs versus other categories, lower margins, and lower public market comparable company multiples, which drive private valuations lower.

Audience companies (the product is free to use for consumers, the company makes money thru ads or leads) drive the second-most value on our list (27%); they are 17% of companies on our list, down from 28% last time; have raised $352m; and are at a 16x multiple of value over capital raised on average.

Enterprise software companies (where a business customer pays for larger scale software, often 'on premises' vs cloud-based; or hardware with software) have raised $268 million on average, and are down in "market share" of our list to 17% from 26% of companies; they drive 12% the value of our list, and a 17x multiple.

SaaS companies (cloud-based software offered often via a 'freemium' or monthly model) have grown to 31% of our list (versus 18%), and 20% the value of our

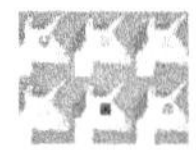

list. They are also among the more capital efficient companies on average in our set, raising $267m on average and are at a 18x return on private capital on average (excluding Veeva).

We've added a new category: Consumer Electronics/Internet of Things, where the consumer pays for a physical product. Five companies make up 6% of list; they have raised $266 million on average and are valued at 18x private capital raised.

An important note – 32% of our set has characteristics of broad or local network effects, where the value of the product/service gets better the more people are part of the system.

6) It has taken ~7 years on average before a "liquidity event" for the 39% who have 'exited' – not including the 61% of our list that is still private. The capital efficiency of these "private unicorns" is surprisingly low, which will likely impact future returns for founders, investors and employees.

When starting a company, many founders may also not realize the journey ahead is more like an ultra-marathon than just a race. It took 6.7 years on average for 33 companies on our list to go public or be acquired (excluding outliers acquired within two years

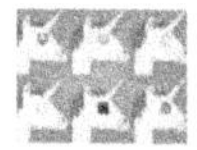

of founding – congrats Instagram, OculusVR, YouTube – you are outliers of the outliers!). Enterprise companies take one year longer to a "liquidity event" vs consumer companies. It's a long journey, well beyond traditional vesting periods.

Just 19 have gone public (23%). The average public market valuation is $8.9 billion. These companies went public after six private rounds of funding, and $329 million in private capital, delivering a 20x at today's valuations (excluding Veeva and Fitbit) versus private capital raised.

14 have been acquired (17%). The sweet spot for an acquisition tends to be at 'lower' valuations – $1.5 billion on average, delivering ~16x on $102m in private capital raised (excluding WhatsApp and YouTube) on average.

Given the valuation premium for private companies today and the increased overhead and quarter-to-quarter pressures of being public, staying private is clearly the preferred option in this market. 51 companies on our list are private; they've raised $516m in 6+ funding rounds to date on average, an astonishing 103% increase (vs $254) vs. our last analysis (for context, in the 'good ol days', Amazon

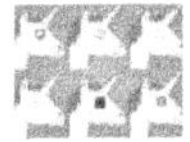

raised $8m before going public, and Google raised ~$26m).

Importantly, valuation vs private capital raised for these private companies is only 8x on average. Unless these companies 'grow beyond' their valuations at exit, this will likely drive lower than historical profits at liquidity for founders, employees and investors.

When will this happen? Given how much capital our private companies have raised in the past few years, most likely have cash to fund 2-4+ more years of runway. If private capital is no longer available in the future, these companies will seek a public offering or acquisition. Some will demonstrate strategically justifiable metrics and have fantastic 'up round' exits; others may see liquidation preferences kick in which will negatively impact founders and employees; others may fulfill the adage "IPO is the new down round", which has been the case for more than half of the public companies on our list. Or worse, some may become "Unicorpses" :)).

The reduction in private company multiples is also a reflection on how venture capital has changed. Ten years ago, the best investors were praised for achieving a 20x return on their $15 million investment = a $300 million return. Many venture investors and

their LPs now invest later, which is perceived to be less risky – and hope to achieve their $300 million by getting a 6x return on a $50 million later-stage investment.

7) Take heart, 'old people' of Silicon Valley – companies with clear product visions, and well-educated, tech-savvy teams of thirty-somethings with history together have built the most successes; 20-something founders, changing CEOs, and "big pivots" are a minority.

The companies in our set were generally not founded by inexperienced, high-school dropouts. The average age at founding was 34 years old (same as our last post). Audience-based company founders were 30 at founding; e-commerce founders were 32; SaaS founders were 35; CE/IoT founders were 36; and enterprise founders were 39.

To note, the founders of our 10 most valuable consumer companies were 29 on average when they founded their companies; and there are a number of young founders (< 25 years old at founding) in our set: Airbnb, Automattic, Box, DropBox, Lyft, Snapchat, Tumblr. But the founders of the two most valuable enterprise companies were 45 years old on average.

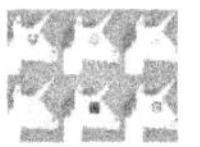

Teams win: a supermajority of companies (86%) has co-founders: 2.6 on average. 85% of co-founders had history together – from school, work or being roommates, the majority having worked together previously.

If at first you don't succeed...76% of companies have founders with entrepreneurial history and a track record of founding something else previously.

Only 12 companies have a sole founder, and none in the top 15 on our list. Unlike in our last analysis where all four sole-founded companies had liquidity events, only two of these companies (New Relic and Tumblr) have had exits.

The overwhelming majority of companies (92%) start with a technical cofounder, and 90% have a founder with experience working in a tech company. It's extremely rare for one of these companies to be started by someone who hasn't worked in tech before. The few companies whose founders had no prior experience working in a tech are largely consumer-oriented companies, like Beats Electronics and Warby Parker.

Education seems kind of important. About half our list has extremely well educated co-founders who are

graduates of a “top 10” U.S. school3; but 19% also have a co-founder who dropped out of college.
Most founding CEOs are scaling through the journey: 74% of companies are still led by their founding CEO, or were led by the CEO through a liquidity event. This says a LOT about the talent of these founding CEOs to scale from seed stage through multiple financings, leadership team changes, hundreds or thousands of team members, and in many cases global expansion, to build the most successful companies of the past decade.
26% of companies have made a CEO change along the way (versus 31% in our last post). Enterprise companies have a higher rate of changing CEO: 32% of enterprise, versus 22% of consumer companies.
83% of companies are working on their original product vision; only 17% significantly changed product focus in a big pivot.
Consumer pivots are more prevalent than in enterprise. In our last analysis, there were just four (or 10%) companies who ‘pivoted’ from their original product vision; and all were consumer companies. The “pivot club” now includes enterprise companies like Slack and MongoDB; and consumer companies like FanDuel, Lyft, Nextdoor, Wish and Twitch.

8) San Francisco maintains its dominance as the new epicenter of the most valuable tech companies; cities like NYC and L.A. are growing in importance.

SF is home to 40% of the companies in our set (vs 38% in our last post); the SF Peninsula is home to 23%; and the East Bay to 4%, for a total of 67% of companies in the Bay Area (versus 69%)
The Big Apple (NYC) is the second-most important geography, home to 12 (14%) of our list, up from 8%.
LA in the house. Los Angeles now has six in our set: Snapchat, Beats Electronics, OculusVR, TrueCar, JustFab, and the Honest Company.
Boston, Austin and Seattle are additional hubs with 3, 2 and 2 companies, respectively.
9) Immigrants play a huge role in the founding and value creation of today's tech companies.

From what we can determine, about 50% of our list has at least one co-founder born in another country. These are remarkable people who in many cases spoke a different language and went to school elsewhere for formative years, then helped create billions in value here in the US. We're grateful these founders and/or their families figured out how to

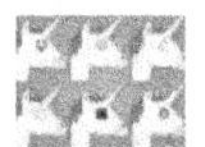

enter and work in our country – and we wonder how many more jobs and how much more value might be created if it were easier for others with outstanding technical and startup skills to get visas to work here. (In case you were wondering: the most common countries that founders came from start with I: India, Iran, Ireland and Israel, in addition to our wonderful neighbor Canada)

10) There's still too little diversity at the top in 2015, but there is movement in a positive direction on gender.

On our last list, there were no female CEOs. So we welcome the two companies on our list with female CEOs: Houzz and Gilt Groupe. And the 10% of companies with female co-founders (up from 5%): CloudFlare, EventBrite, FanDuel, Gilt Groupe, Houzz, NextDoor, Kabam, and The Honest Company. So while 2.4% of CEOs is little to celebrate, this is an improvement from zero.

It is not easy to figure out from publicly available information, but we estimate about 30% of companies in our set have no females on the leadership team. The majority of female senior leaders we could identify are in CFO, VP HR, GC, Sales and CMO roles;

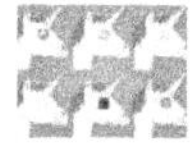

we could only find a few companies with female leaders in product or engineering, which seems like a great opportunity for progress. And Kudos to Gilt Groupe, Lending Club, New Relic, and ZenDesk who seem to have among the most gender diverse teams in our set.

From what we can tell, ~70% of companies in our set have no gender diversity at the board level. This also seems a huge opportunity to improve outcomes and send an important message from the top. We won't call out the companies with no diversity on their leadership teams or boards – but they exist, and we hope they are paying attention to this important driver of outcomes and culture. Efforts like Sukhinder Singh Cassidy's BoardList will help identify great candidates for these boards.

To note: we aren't able to track racial or ethnic diversity as well as we need to report; this is an important field we hope to track in a future analysis.

decacorn2

So, what does this all mean?

The most striking takeaway for us from this analysis is the growth of "paper unicorns" and their surprisingly

low capital efficiency. While we believe some increase is due to fantastic market fundamentals, much seems due to the low incentive to trade publicly, a fiercely competitive environment, and private capital flocking to growth that has caused companies to focus on 'getting big fast', and has also pushed valuations out of whack with public markets.

Because many investors have protection through preferred stock, and many founders "take some off the table" in later stage rounds, when paper unicorns become public or acquired unicorns, non-founder employees will likely feel the most pain and disappointment if there is a negative gap between the exit value and today's private market share prices.

(Which makes us think: While being a unicorn is cool, you know what's really cool for founders, employees and investors? Making a 20x+ multiple on private capital raised when your super awesome company has an exit.)

There are also many consistent lessons with our last analysis. These companies were largely founded by co-founder teams with clear product visions, history

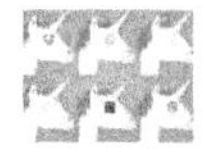

together, experience working in tech, track records of entrepreneurship, and with a technical founder on the team. All have opportunities to improve their outcomes and cultures by adding diversity to their teams and boards of directors.

And, they were founded by committed leaders who are on a path to scale their businesses for a decade or more. The probability a founder could start with a scrappy dream, then develop the skills to lead through hundreds of product shipments, ups and downs, serial fundraisings and thousands of employees while maintaining the faith of their teams, investors and boards – seems quite unlikely.

It's really remarkable and this whole analysis could be an ode to these special founders and teams who are going the distance to achieve the improbable.

So we continue to tip our hats to these 84 companies who delight millions of customers with fantastic products, are outstanding at fundraising, and recruit and retain team members in the most brutal recruiting environment we've known.

They are the lucky/tenacious/genius few of the Unicorn Club, and we look forward to learning more about them, and their future compatriots. https://techcrunch.com/2015/07/18/welcome-to-the-unicorn-club-2015-learning-from-billion-dollar-companies/

If money raised doesn't predict long-term value creation, what does? The research points to two interesting correlations. The first is the age of the company at IPO. "Companies that go public between the ages of six and 10 years generate 95% of all value created post-IPO," Ramadan says.

They found that the vast majority of post-IPO value creation comes from companies they call "category kings," which are carving out entirely new niches; think of Facebook, LinkedIn, and Tableau. Those niches are largely "winner take all"—the category kings capture 76% of the market.

Tech start-ups are in a race to define new product categories, and the pace has quickened.

Chapter 2: Concept

1. IDENTIFY MARKET SIZE (REVENUES PER YEAR):
2. IDENTIFY TOTAL ADDRESSABLE MARKET SIZE (REVENUES PER YEAR):

To build a unicorn, both of these should be in the billions. You need to be able to show that the company can capture $100M in recurring revenues in years 5 to 10. This means that you have built a unicorn.

1. IDENTIFY UNIT ECONOMICS (PROFITABILITY) OF THE COMPANY:
    a. IDENTIFY COST OF ACQUISITION PER CUSTOMER (CAC):
    b. IDENTIFY COST PER USER PER MONTH:

c. IDENTIFY REVENUE PER USER PER MONTH:
d. IDENTIFY LIFETIME VALUE PER USER (LTV):

To build a unicorn, the REVENUE PER USER PER MONTH SHOULD BE 30% to 50% higher than COST PER USER PER MONTH. This means that when you scale, you do so extremely profitably. These sort of margins are what billion-dollar consumer/internet companies get including Facebook and Linkedin.

Size of market means how many revenues per year is generated by that industry. You want to achieve $100M recurring revenue per year of that market in 7 years. What % is that? This means a billion dollar company.

Billion Dollar Markets that are (y) Ripe for Disruption checks the box. If you're at a $2.9B TAM, and that's truly an addressable market ... that works. Check.

The tactical question though is can you get to $100m in ARR in 6-7 years if everything goes really well?

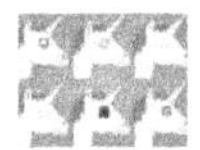

Even if your nominal TAM is high, my #1 recommendation if you are pre-MSP (Minimum Sellable Product) is to do your own model to get to $100m in 7 years.

Any product that can really sell at a $100m ARR in 7 years has a billion+ TAM, by definition.

that TAM needs to be very large. At least a Billion.

He said to us that he was only interested in markets where the opportunity was in billions, not millions.

For example, using some rough math, lets say a software startup targets a $1bn market opportunity. Let's assume that in a good outcome the startup is able to achieve 20% market penetration 6 years out, so $200m in annual revenues. Software operating profit margins might fall in the 20% - 35% range. Lets' assume 25% operating profit margins, so the startup is generating $50m in EBIT 6 years out. Assuming that the market growth rate has stabilized at 10% per year, this implies that the value of the company is approximately 8x-10x EBIT or $500 million at the higher end of the range. If over the life of the

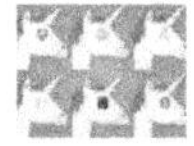

company the VC invested $25 million and owns 25% of the equity, then the value at exit in this good outcome is $125m or a 5x return and an $100m gain. $100m is 10% of the $1bn fund size and so would "move the needle".

A general rule is that most VCs want to believe that their investment has a reasonable likelihood of a 10x return. To make this estimate, the VCs start at your exit and work backward.

In order to make educated guesses, they will need to understand the amount of capital you will require over time, the relative values you will raise such capital at and the approximate size of your exit.

Do this math ahead of time - it will help you select the right firms to target. Some will want to put in $20M over your life, others only a few million. However, they will all want out-sized (10x) returns.

Market size is not an explicit variable in the above calculation, although it is clearly an influencer, as your share of the market will dictate your cash flows and ultimate valuation.

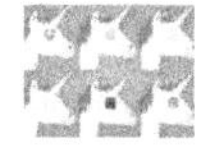

If a VC puts in $5M and you ultimately achieve $30M of recurring revenue, which might represent 10% - 20% of the total market, that VC will be happy. Applying a conservative multiple range, your exit would be $150M - $200M...which is a crap load of Chiclets.

it talks about "addressable market" which is more important than market size.

Step 2: Find $1,000,000 worth of customers.

Now that you've found an idea, it's time to assess whether there's a big enough pool of prospective buyers. In this step, you'll also want to ensure your market isn't shrinking, and that it fares well compared to similar markets.

I use Google Trends, Google Insights, and Facebook ads when I'm in this part of the process. They're great tools that help me evaluate the growth potential of my target market.

For example, let's say you decide to build information products for owners of Chihuahuas (remember "Yo quiero Taco Bell"?). Here's how I would check to see if there are enough customers:

1. Search Google Trends for the term "chihuahua" and other similar words (e.g. poodle, dogs) for comparison:

(Click image to expand)

We can see that the word "chihuahua" has a decent search volume (relative to "dogs"), and that "poodle" isn't as popular. It also looks like the number of searches for "chihuahua" has been relatively stable for the last few years.

2. Double-check on Google insights:

Google Insights is great, because it breaks down the search data by location (i.e. what regions the searches are coming from), by date, and what they're searching

for (news, images, products). Click here to see the full report for the above chart.

3. Look at the total number of people available on Facebook for dogs:

3.1 million. Not bad, not bad.

And for Chihuahuas:

84,260 people. Score.

You can also see if there is a large property that you can piggyback on.

Paypal did this with eBay, AirBnb is doing it with Craigslist home listings, and AppSumo looks to the 100 million LinkedIn users. If you can find a comparable site with a large number of potential customers, you'll be in good shape.

Step 3: Assess your customer's value.

Once you've found your idea and a big pool of potential customers, you'll need to calculate the value of those customers. For our example above, we'll need to estimate how much a Chihuahua owner (i.e. our customer) is worth to us. This will help us determine the likelihood of them actually buying our product, and will also help with pricing. Here's how we do that:

1. Find out how much it costs, on average, to buy a Chihuahua (about $650). This is the base cost.

2. See how much it costs to maintain a Chihuahua each year (i.e. recurring costs). Looks like it's between $500-3,000. For this example, we'll call it $1,000.

3. Look up their life expectancy, which is roughly 15 years. This is the number of times they'll have to pay those recurring costs.

Therefore, a Chihuahua's average total cost of ownership is:

[$650 + ($1,000*15)] = $15,650

Damn... you could buy a lot of burritos with that kind of cash. Silly dog owners.

In any case, these owners are already committing to spend a LOT of money on their dogs (i.e. they are valuable). After putting down $650 on the dog itself and an average of $80/month on maintenance (a.k.a. food), spending $50 on an information product that could help them train their Chihuahua–or save money, or create a better relationship between them, etc.–does not seem unreasonable. Of course, the product doesn't have to cost $50, but we now have some perspective for later deciding on a price.

Now we need to utilize the TAM formula (a.k.a. Total Available Market formula), which will help us see our product's potential to generate a million dollars.

Here's the TAM formula for estimating your idea's potential:

(Number of available customers) x (Value of each customer) = TAM

If TAM > $1,000,000, then you can start your business.

Let's plug in some basic numbers to see the TAM for our Chihuahua information product:

(84,260 available customers) x ($50 information product) = $4,213,000

We have a winner!

Okay, obviously you are not going to reach 100% market penetration, but consider the following...

1. This is only through Facebook traffic.

2. This does not include the 5,000,000 monthly searches for "Chihuahua" on Google:

3. This is only for one breed of dog. If you find success with Chihuahuas, you can easily repeat the process many times with other dog breeds.

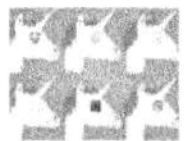

4. This is only for one product. It's far easier to sell to an existing customer than it is to acquire new ones, so once we've built up a decent customer base, we can make even more products to sell to them.

By all measures, it appears that we have a million dollar idea on our hands. Now we can move on to the final step!

Step 4: Validate your idea.

By now, you have successfully verified that your idea has that special million-dollar-potential. Feels good, right? Well, brace yourself — it's time to test whether people will actually spend money on your product. In other words, is it truly commercially viable?

This step is critical. A lot of your ideas will seem great in theory, but you'll never know if they're going to work until you actually test your target market's willingness to pay.

For instance, I believed AppSumo's model would work just on gut-feeling alone, but I wasn't 100% convinced people wanted to buy digital goods on a time-limited

basis. I mean, how often do people find themselves needing a productivity tool (compared with, for instance, how often they need to eat)?

I decided to validate AppSumo's model by finding a guaranteed product I could sell, one with its own traffic source (i.e. customers).

Because I'm a frequent Redditor and I knew they had an affordable advertising system (in addition to 3 million+ monthly users), I wanted to find a digital good that I could advertise on their site. I noticed Imgur.com was the most popular tool on Reddit for sharing images, and they offered a paid pro account option ($25/year). It was the perfect fit for my test run.

I cold-emailed the founder of Imgur, Alan Schaaf, and said that I wanted to bring him paying customers and would pay Imgur for each one. Alan is a great guy, and the idea of getting paid to receive more customers was not a tough sell☺ The stage was set!

Before we started the ad campaign, I set a personal validation goal for 100 sales, which would encourage

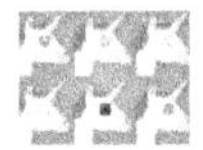

me to keep going or figure out what was wrong with our model. I decided on "100" after looking at my time value of money. If I could arrange a deal in two hours (find, secure, and launch), I wanted to have a return of at least $300 for those two hours of work. 100 sales ($3 commission per sale) was that amount.

By the end of the campaign, we had sold more than 200 Imgur pro accounts. AppSumo.com was born.

I share this story because it illustrates an important point: You need to make small calculated bets on your ideas in order to validate them. Validation is absolutely essential for saving time and money, which will ultimately allow you to test as many of your ideas as possible.

Here are a couple methods for rapidly validating whether people will buy your product or not:

Drive traffic to a basic sales page. This is the method Tim advocates in The 4-Hour Workweek. All you need to do is set up a sales page using Unbounce or WordPress, create a few ads to run on Google and/or Facebook, then evaluate your conversion rate for ad-

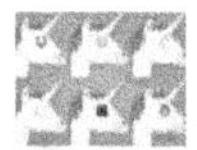

clicks and collecting email addresses. This is how we launched Mint.com (see one of our original sales pages here). You are not looking for people to buy; you are simply gauging interest and gathering data.

[Note: With Facebook advertising, $100 can get you roughly 100,000 people viewing your ad, and about 80 people visiting your site and potentially giving you their email addresses.]

Email 10 people you know who would want your pseudo-product, then ask them to send payment via Paypal. This might sound a bit crazy, but you're doing it to see what the overall response is like. If a few of them send payment, great! You now have validation and can build the product (or you can refund your friends and buy them all tacos for playing along). If they don't bite, figure out why they don't want your product. Again, the goal is to get validation for your product, not to rip off your friends.

Of course, there are other techniques for validating your product (like Stephen Key leaving his guitar pick designs in a convenience store to see if people would try to buy them). However, I've found these two

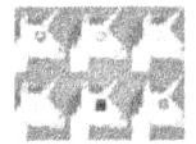

methods to be super efficient and effective for validating ideas online.

No need to get fancy if it does the trick.

The Final Frontier: Killing Your Inner Wantrepreneur

We made it! You officially have a $1,000,000 idea on your hands and you know for a fact that people are willing to pay for it. Now you can get started on actually building the product, creating your business, and freeing yourself from the rat race!

I can just see it... You're all nodding and thinking, "Hey, this Noah guy is pretty snazzy!" (Sorry ladies, I'm taken.)

So, what now?

- You are inspired. Check.
- You want to do something. Check.
- You get a link to a funny YouTube video, then you open up Reddit. Check.
- Suddenly, everything you thought you were going to do goes down the drain. Check.

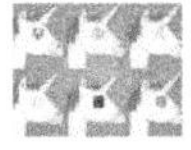

– You and I softly weep. Check.

I want to challenge you! Whoever generates the most profit (not just revenue) within 14 days of this article will win some fantastic goodies. First, here are the basic rules and the process:

– Contest void where prohibited.
– The business/product must be new. This means either a landing page created from scratch using Unbounce or WordPress above.
– Results and proof of some type must be submitted as a comment below no later than 1am PST Saturday on October 8, 2011. Don't cut it too close; if a timezone misjudgment knocks you out, we can't make exceptions.
– Put your 14-day profit number (or increase) in the FIRST line of your comment.
– Ultimately, verifiable proof with lower number beats unverifiable proof with higher number.

[NOTE: THIS CONTEST HAS ENDED. Still need help starting a business? Check out AppSumo's "How to Make your First Dollar" course.]

The prizes:

– $1,000 credit from AppSumo.com
– Roundtrip flights to Austin, Texas to have the most delicious tacos in the world with Noah Kagan, CEO of AppSumo. Sorry, but we can only cover flights within the USA. If you want to hoof it to the US, we can then pick up from there.
– Above all: your $1,000,000 business, of course!

Don't let this post become another feather in your Wantrepreneurship cap. Just follow the steps and start working towards your $1,000,000 business! Remember, you can start laying the foundation for your product without building anything.

All you need is one weekend.

http://fourhourworkweek.com/2011/09/24/how-to-create-a-million-dollar-business-this-weekend-examples-appsumo-mint-chihuahuas/

Customer acquisition cost (CAC) is a metric that has been growing in use, along with the emergence of

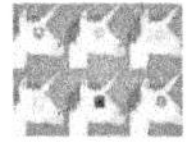

Internet companies and web-based advertising campaigns that can be tracked.

Traditionally, a company had to engage in shotgun style advertising and find methods to track consumers through the decision-making process.

Today, many web-based companies can engage in highly targeted campaigns and track consumers as they progress from interested leads to long-lasting loyal customers. In this environment, the CAC metric is used by both companies and investors.

CAC, as you probably know, is the cost of convincing a potential customer to buy a product or service. In this article, we will explain the CAC metric in more detail, how you can measure it, and what steps you can take to improve it.

What the CAC Metric Means to You

As mentioned above, the CAC metric is important to two parties: companies and investors. The first party includes outside, early stage investors who use it to analyze the scalability of new Internet technology companies. They can determine a company's

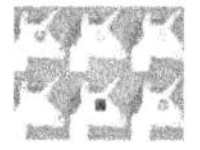

profitability by looking at the difference between how much money can be extracted from customers and the costs of extracting it.

For example, in terms of the upstream oil market, if an oil supply is in an area requiring heavy infrastructure investments, the amount applied to extract the oil may be greater than its market price per barrel.

Investors view Internet-based companies through the same lens. They are concerned with the current relationship, not on future promises of improving the metric, unless they can be justified.

The other party interested in the metric is an internal operations or marketing specialist. They use it to optimize the return on their advertising investments. In other words, if the costs to extract money from customers can be reduced, the company's profit margin improves and it makes a larger profit.

Then, investors are more interested in providing the company with the resources it needs, partners are more committed to growth, and the company can use

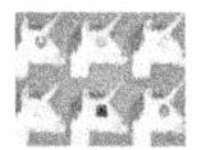

the improved profit margins to pass the value to its customers for a greater market position.

How You Can Measure CAC

Basically, the CAC can be calculated by simply dividing all the costs spent on acquiring more customers (marketing expenses) by the number of customers acquired in the period the money was spent. For example, if a company spent $100 on marketing in a year and acquired 100 customers in the same year, their CAC is $1.00.

There are caveats about using this metric that you should be aware of when applying it. For instance, a company may have made investments on marketing in a new region or early stage SEO that it does not expect to see results from until a later period. While these instances are rare, it may cloud the relationship when calculating the CAC.

It is suggested that you perform multiple variations to account for these situations. However, we will provide some examples of calculating the CAC metric in its most pragmatic and simple form with two examples.

The first company (Example 1) has a poor metric. The second (Example 2) has a great one.

Example 1: An ecommerce company
In this example, we take a fictitious ecommerce company that sells organic food products. The company spent $100,000 on advertising last month, and its marketing team says 10,000 new orders were placed. This suggests a CAC of $10, a figure that has no meaning in itself.

If a Mercedes-Benz dealer has a CAC of $10, the management team will be delighted when looking at the year's financial statements.

However, in the case of this company, the average order placed by customers is $25.00, and it has a markup of 100% on all products. This means that on average, the company makes $12.50 per sale and generates $2.50 from each customer to pay for salaries, webhosting, office space, and other general expenses.

While this is the quick and dirty calculation, what happens if customers make more than one purchase

over their lifetime? What if they completely stop shopping at brick and mortar grocery stores and buy from only this company?

The purpose of customer lifetime value (CLV) is specifically designed to resolve this. You can find a CLV calculator by simply searching in your favorite search engine. In general, this metric helps you form a more accurate understanding of what the customer acquisition cost means to your company.

A $10.00 customer acquisition cost may be quite low if customers make a $25.00 purchase every week for 20 years! However, in this ecommerce company, they are struggling to keep customers and most of the customers make only one purchase.

Knowing the CAC for each of your marketing channels is what most marketers want to know. If you know which channels have the lowest CAC, you know where to double down on your marketing spend. The more you can allocate your marketing budget into lower CAC channels, the more customers you can obtain for a fixed budget amount.

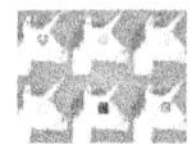

The simple approach is to break out your spreadsheet and gather all your marketing receipts for the year, quarter or month (however you want to do it) – and add up those amounts by channel. For example, how much did you spend on Google Adwords and Facebook advertising? In this case you might put this in a column called "PPC" or "Pay-Per-Click". How much did you spend on SEO and blogging? This might go into a column called "Inbound Marketing Costs".

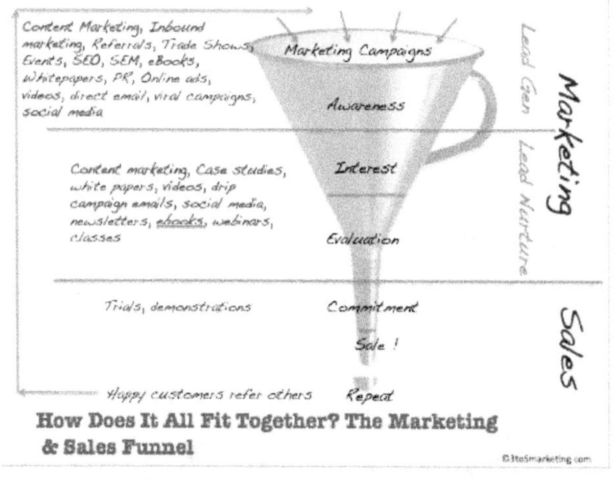

How Does It All Fit Together? The Marketing & Sales Funnel

When someone on your website does something you want them to do (i.e., sign up, make a purchase, fill out a form, etc.), it is known as a conversion.

A funnel is used to track the steps that lead up to that conversion. For example, Ecommerce companies want people to purchase products on their website. Their funnel may have these steps – visited site > viewed product > placed product in cart > purchased.
Using a funnel report you can see where people are dropping off in the path to conversion.
Both Google Analytics and Kissmetrics provide funnels. Each have their unique use cases. Kissmetrics provides additional reports in addition to the Funnel Report.

The Top 5 Kissmetrics Reports Every SaaS Marketer Needs

Today's SaaS marketers face many challenges. Their job, of course, is to spread awareness and drive customer acquisition.

On top of these difficult tasks, marketers are always under pressure to improve their numbers and performance. You've got a 3% conversion rate? Great, how do you double that in the next month?

Kissmetrics is here to help marketers. Our reports enable marketers to track and analyze their online marketing campaigns. Today's post describes five Kissmetrics reports every SaaS marketer needs.

These reports will help you track your marketing performance and provide you with guidance on what is holding back your growth.

1) Signup Funnel – Increase Conversions by Finding Where Visitors Drop Off
If you're not using a signup funnel, you're really missing out on very important data.

Every SaaS product has a set of steps visitors need to go through before they can become customers. Here are the typical steps:

Visitors need to visit the site
They need to sign up for a trial
They need to start using the product
They need to upgrade to paying once the trial is complete

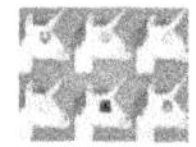

Visitors can take actions in between the steps (such as visiting pricing page, about page, etc.), but those actions are not necessary to get to the next step.

A Funnel Report helps marketers identify where visitors are dropping off in their funnel. Here's how a funnel looks:

saas signup funnel kissmetrics report

This shows us that 3% of visitors convert to signing up for a trial. To improve on that, we can do some data digging and try to get more traffic to our site from the channels that deliver the most signups. We can also run some A/B tests to determine which offers and calls-to-action attract the most people to sign up.

Click here to view a demo of the Funnel Report.

2) A/B Test Report – See How an A/B Test Impacts Your Entire Funnel

Kissmetrics has an awesome report called the A/B Test Report. It allows you to run an A/B Test and see how the variants impact your entire funnel. You can still set up your tests in Optimizely, VWO, or whatever tool

you prefer. But after you set them up, use Kissmetrics to see how each variant page impacts your funnel.

It's really easy. All you have to do is pick your conversion event and the experiment.

saas signup funnel kissmetrics report

The conversion event is an action you want your visitor to take. You can pick any part of your funnel, or any part outside your funnel. As long as you're tracking it, you can test for that outcome.

Want to see if a homepage headline test moved the needle at the bottom line? No problem, just set the bottom-line conversion (for example "Billed") as your conversion event, and get your answer in less than 10 seconds.

The A/B Test Report can also save you from false positives. You may run a test or a variant that gets more free trial signups, but loses further on down the funnel, where it matters more. If you were not using the A/B Test Report, you would launch a losing variant

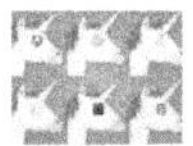

to all your visitors, and it would actually hurt your business!

Click here to view a demo of the A/B Test Report.

3) Cohort Report – Track Acquisition and Retention Performance
A Cohort Report puts people into groups and tracks their behavior over time. This makes it very useful for tracking acquisition and retention. If you're in SaaS, you need to track login retention with the Kissmetrics Cohort Report. This will help you compare your marketing campaigns and see which campaign or promo had the best effect on retention.

Don't stop at just sign-up rates and call it a day. Look at how engaged your users are after they sign up. A login retention cohort report can do that for you.

Here's how it looks in Kissmetrics:

cohort retention kissmetrics

On the left side, we see each marketing channel and the number of people from that channel who have

logged in. On the right, we see the retention we get from each marketing channel. The higher the number, the better the retention. The better the retention we get from a specific channel, the more money/effort we should funnel into acquiring customers from that channel.

Click here to view a demo of the Cohort Report.

4) People Search – Find Signed Up But Unengaged Customers
As a marketer, you need to deliver more than signups. You need to deliver customers. Signups are great, but if they don't use your product, they are no more useful than people who do not sign up. You have no chance of being paid for your product. To help you find these unengaged free trial users, you can use the Kissmetrics People Search.

The People Search lets you find users who share a certain behavior. In our case, we'll want to find the people who have signed up but not logged in. Here's how our criteria is set in People Search:

cohort retention kissmetrics

This will show us the people who fit this criteria in the past 7 days. We'll add more data to our search by requesting when they signed up. We'll click Search and get our list:

cohort retention kissmetrics

And we get a list of people who have signed up but not logged in.

We need to figure out why these people aren't using our product. They've shown interest by signing up, but they aren't using it. The best next step is to email each of them a personal email and find out why they haven't logged in.

You aren't limited to using login as your engagement metric. As long as you're tracking an action/event, you can find the people. Here are a few other ideas:

People who have signed up, logged in, but not used a feature
People who have signed up, logged in, but not taken some key action you want them to take

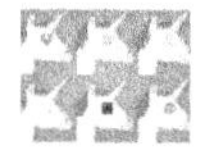

People who have signed up, logged in, and viewed a support article (these people need help)
People who have signed up, logged in, and submitted multiple support tickets
Again, as long as you're tracking an action/event, you can find the people who have, and have not, taken/triggered the action/event.

Click here to see how People Search works.

5) Revenue Report – Know Where Your Most Valuable Customers Are Coming From
The Kissmetrics Revenue Report is just what it sounds like. It reports your revenue, giving you various metrics that can help you build a more efficient marketing operation.

Viewed as a whole, your revenue is just one number, much like your traffic. You can't get a lot of insight from looking at a number. With the Kissmetrics Revenue Report, you can break up your revenue into groups (called segments) based on anything you wish. If you're a marketer, you'd segment by marketing channel. Here's how it looks:

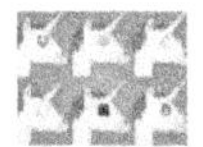

cohort retention kissmetrics

We're looking at the "first ever" marketing channel. This basically tells Kissmetrics to segment people by their first channel. So if a person first came to your site via an organic search, they'd be put in the Organic segment. If they came from a social site, they'd be put in the Social segment. The None segment is for people who do not meet any criteria for the other channels.

This data shows you how each of your marketing channels are performing. A channel may perform better at acquisition, but if you're not using the Revenue Report to track how valuable customers from each channel are overall, you'll be endlessly spending money on a leaky bucket. Your job doesn't stop at signups.

When you know where your most valuable customers are coming from, you can put more of your effort and money into acquiring users from those channels.

You aren't limited to segmenting revenue by marketing channels. Here are a few other ideas (which you would need to be tracking):

Subscription plan type
Referring URL
If you're using UTM's, Kissmetrics will automatically pick those up, and you can segment revenue by any parameter

Discount rate in your models changes through time; update as progress through database acquisition, self-advocacy, activation, utilization, monetization, scale (derisking linked to execution of Work Breakdown Schedule).

Seed: pre-database MVP: 80% discount rate
Angel: self-advocacy & activation: 50%-70% discount rate
Series A: utilization & monetization: 40%-60% discount rate
Series B: monetization & scale: 30%-50% discount rate
Growth: Scale & new markets: 20%-40% discount rate
Mezzanine: Scale in old & new markets: 15%-30% discount rate

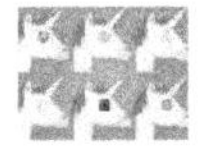

IPO: Scale globally: 7-15% discount rate
Post-IPO Minority Ownership: Mature & New Product Development: 3.5%-10% discount rate

http://people.stern.nyu.edu/adamodar/New_Home_Page/datafile/wacc.htm

Net margins by industry:
http://pages.stern.nyu.edu/~adamodar/New_Home_Page/datafile/margin.html

Revenue growth rates by industry:
http://pages.stern.nyu.edu/~adamodar/New_Home_Page/datafile/histgr.html

Unit economics (profitability):
Consumer/internet cost per user per month: $1
Revenue per user per month:
Cost Aqcuisition per customer (CAC):
Liftime value per user (LTV):

So like a multi-factor model for wealth maximization in an entrepreneurial context. Factors would include the choice of industry which are primarily going to be internet/technology or life sciences. Another factor is

% recurring revenue. Another growth rate in users or revenue. Final factor could be market dominant or perceived throw at to other strategics. When another strategic seed you as a legit threat you get a premium valuation; this was the case for Whatsapp and Linkedin.

Strategics in their models run disruptive scenarios; you can determine a factor or maybe a threshold called disruption premium. The amount over a normal valuation premium in a non-disruptive context

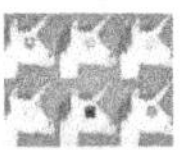

Chapter 3: Platform

1. MODEL PLATFORM ARCHITECTURE IN TERMS OF BASE PHP PLATFORM AND MODS ONTO IT

Chapter 4: Database(s)

1. BUILD DATABASE OF 25 HIGHEST PRIORITY USERS OF PLATFORM AND COLLECT EMAIL, NAME, NUMBER

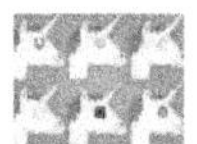

Chapter 5: Self-Advocacy

1. SELF-ADVOCATE TO THE 25 HIGHEST PRIORITY USERS OF PLATFORM AND NOTIFY THEM THAT YOU WILL BE SENDING THEM VALUE AND THAT THIS IS THEIR ACCOUNT TO CLAIM IT GIVING THEM LOGIN CREDENTIALS

Chapter 4: Activation

1. SCREEN SHARE WALK THROUGH WITH THE HIGH PRIORITY USER AND SHOW THEM HOW IT WORKS AND GET THEM TO LOGIN BY THEMSELVES

Chapter 5: Utilization

1. USER UTILIZES PLATFORM BY THEMSELVES FOR A TRIAL PERIOD OF TIME (3 TO 6 MONTHS)

Chapter 6: Monetization

1. SEND USER THE INVOICE FOR USING THE PLATFORM. FEE SHOULD BE MEMBERSHIP FEE BASED UPON STANDARD MARKET FEE.

Chapter 9: Scale

1. BUILD DATABASE OF 26-50th HIGHEST PRIORITY USERS OF PLATFORM AND COLLECT EMAIL, NAME, NUMBER

Each stage of growth:

Prove the idea: $0 to $1 million
With the right idea, you can generally raise $1 million with a seed round. You don't have to bet the whole company on an A round. At Zuora, for example, our idea was that subscription-based companies shouldn't have to build their own billing systems just because

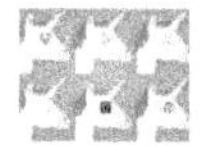

telco billing systems were too expensive and regular enterprise resource planning (ERP) systems didn't provide the required features. So we built a prototype and spoke with over 50 potential customers. Through Salesforce.com, we put screen shots on a website, invited people to participate in a beta, showed a demo to everyone interested, and asked what they thought. We had 200 customers lined up when we launched.

Prove the product: $1 million to $3 million

What product do your customers really need? You may have to cull your ideas and decide what to build first and what can be pushed out to later years. Or you may have to do the opposite. We initially thought a simple billing system was sufficient. Then we realized we needed a very broad platform, including application program interfaces (APIs), payment systems, commerce platforms, and tax engines.

You're still learning at this stage, but you have to complete the process now. You won't have time to dramatically re-engineer your product during the next stage because you'll be looking to scale. You'll also need a coherent product roadmap in place.

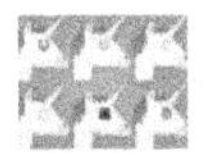

Prove the market: $3 million to $10 million
This stage is very exciting. You're hiring salespeople, spending money on marketing, and focusing on execution. Defining your market is now essential. Is it truly a billion dollar market? $500 million? $100 million? For your seed or A round, it was all about hopes and dreams. Now investors will demand to know how big this market really is, and you'll need the story and statistics to prove it.

When we started Zuora, many assumed our product was strictly for the software as a service (SaaS) industry. We knew any company could benefit from our solution, so we gathered the proof points by going to media, financial services, and healthcare companies. We even went to a public company. Don't automatically limit yourself to focusing on a single vertical. It can actually box you into a smaller market.

Prove the business model: $10 million to $30 million
Now it's time to focus on your business model. What is your customer acquisition cost? Your churn rate? How much does your R&D really cost? What is your

customer lifetime value? Your gross margins? Do you have a positive net dollar retention? Are the trends getting better?

I'm not suggesting you completely ignore business model metrics during earlier stages, but the cost of a sub-optimal business model at $10 million is small. But now you need to make the tough choices so you don't find yourself at a $30 million run rate with a money-losing business. Do you have the right sales model? If not, swap out the sales team. Are you chasing unprofitable customers? If so, make them profitable or stop doing it. Do your price points allow for strong gross margins? If not, figure out how to increase revenues per customer. Keep in mind that at this stage "chasing anything that moves" is one of the top reasons companies stall.

Prove the vision: $30 million to $100 million
If you're going to take public money or larger investments, your new investors need to know you're a safe bet and will be around for the long haul. They aren't venture investors that expect some of their bets to fail. They want minimal risk. They want the right management team and board. They want predictable

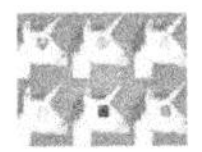

execution and positive cash flow, or at least a clear path to achieving it. They also need to see that the market size is real and that you're a leader with the right set of products, people, and partners to create a sustainable competitive advantage.

Prove the industry: $100 million to $300 million
At this point, your company has scale. Smaller companies are starting to surround you, like planets circling a sun. They like your customer base and marketing. They like how many salespeople you have and hope to use those salespeople to increase their reach.

Now is the time to build an ecosystem. It won't happen overnight, but by the time you reach $300 million, you need to be able to explain how, as the center of an expanding industry, you will continue to grow with it. By creating a 'platform' with a burgeoning ecosystem, you are actually demonstrating how you will become a billion dollar business.

Prove the company: $300 million to $1 billion

You have everything in place to reach a billion. Now it comes down to execution. Never get complacent. Never lose sight of how the industry and competitors are evolving, especially start-ups with new and potentially disruptive ideas.

Few companies make it to a billion, but by breaking the journey into stages, the process becomes more comprehensible and less forbidding. Just recognize that each stage is different and what you are trying to prove is different. Embrace change. Embrace the switchback. For only when you round the turns can you begin to understand John D. Rockefeller's phrase, "Don't be afraid to give up the good to go for the great."

Tien Tzuo is the CEO of Zuora, a software company that designs and sells SaaS applications for companies with a subscription business model.

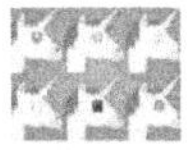

Chapter 11: Exit

1. IDENTIFY 3 TO 5 MOST LIKELY STRATEGIC BUYERS OF YOUR PLATFORM. REACH OUT TO THE PERSONALLY WHEN YOU HAVE $100M IN RECURRING REVENUE AND EXIT AT THE MARKET REVENUE MULTIPLE BASED UPON PRECEDENT TRANSACTIONS.

Exiting at a 10x+ revenue multiple means that you have exited at a valuation of $1 billion dollars and have thus created a unicorn.

Size of market means how many revenues per year is generated by that industry. You want to achieve $100M recurring revenue per year of that market in 7

years. What % is that? This means a billion dollar company.

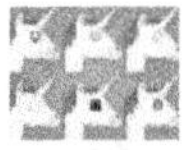

Chapter 11: Cases

M&A Nexus:
Concept: M&A network and marketplace connecting strategic and financial buyers to advisors representing middle market companies for sale.
Market size: $1.4 Trillion (capital marketplace)
Total addressable market: $900 Billion (M&A transactions in non-tech companies)
UNIT ECONOMICS (PROFITABILITY) OF THE COMPANY:
a. IDENTIFY COST OF ACQUISITION PER CUSTOMER (CAC): database cost/# of companies in database = $3000/12,000 = $.25
b. IDENTIFY COST PER USER PER MONTH: server cost/total users = $150/12,000 = $.02
c. IDENTIFY REVENUE PER USER PER MONTH: revenue per year/12 = $2,000/12 = $166.66

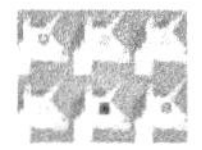

d. IDENTIFY LIFETIME VALUE PER USER (LTV): revenue per year x life expectancy = $2,000 x 60 = $120,000
MODEL PLATFORM ARCHITECTURE IN TERMS OF BASE PHP PLATFORM AND MODS ONTO IT
DATABASE OF 25 HIGHEST PRIORITY USERS OF PLATFORM AND COLLECT EMAIL, NAME, NUMBER. .XLS DATABASE NAME:
SELF-ADVOCATE TO THE 25 HIGHEST PRIORITY USERS OF PLATFORM AND NOTIFY THEM THAT YOU WILL BE SENDING THEM VALUE AND THAT THIS IS THEIR ACCOUNT TO CLAIM IT GIVING THEM LOGIN CREDENTIALS. .DOC MESSAGE FILE NAME:
SCREEN SHARE WALK THROUGH WITH THE HIGH PRIORITY USER AND SHOW THEM HOW IT WORKS AND GET THEM TO LOGIN BY THEMSELVES
USER UTILIZES PLATFORM BY THEMSELVES FOR A TRIAL PERIOD OF TIME (3 TO 6 MONTHS)
SEND USER THE INVOICE FOR USING THE PLATFORM. FEE SHOULD BE MEMBERSHIP FEE BASED UPON STANDARD MARKET FEE.
BUILD DATABASE OF 26-50th HIGHEST PRIORITY USERS OF PLATFORM AND COLLECT EMAIL, NAME, NUMBER

IDENTIFY 3 TO 5 MOST LIKELY STRATEGIC BUYERS OF YOUR PLATFORM. REACH OUT TO THE PERSONALLY WHEN YOU HAVE $100M IN RECURRING REVENUE AND EXIT AT THE MARKET REVENUE MULTIPLE BASED UPON PRECEDENT TRANSACTIONS.
Axial, LinkedIn, Alibaba

DISCOUNT RATE STATUS:
Seed: pre-database MVP: 80% discount rate
Angel: self-advocacy & activation: 50%-70% discount rate
Series A: utilization & monetization: 40%-60% discount rate
Series B: monetization & scale: 30%-50% discount rate
Growth: Scale & new markets: 20%-40% discount rate
Mezzanine: Scale in old & new markets: 15%-30% discount rate
IPO: Scale globally: 7-15% discount rate
Post-IPO Minority Ownership: Mature & New Product Development: 3.5%-10% discount rate

FINANCIAL MODEL & VALUATION (DCF):

M&A Nexus

	2014	2015	2016	2017	2018	2019	2020	2021	2022	2023	2024	2025	2026	2027	2028	2029
Membership (recurring revenue)				$ 20,000.00	$ 60,000.00	$ 180,000.00	$ 270,000.00	$ 405,000.00	$ 607,500.00	$ 911,250.00	$ 911,250.00	$ 911,250.00	$ 911,250.00	$ 911,250.00	$ 911,250.00	$ 911,250.00
Profit per year	**$0.00**	**$0.00**	**$0.00**	**$20,000.00**	**$60,000.00**	**$180,000.00**	**$270,000.00**	**$405,000.00**	**$607,500.00**	**$911,250.00**	**$911,250.00**	**$911,250.00**	**$911,250.00**	**$911,250.00**	**$911,250.00**	**$911,250.00**
Discount Rate	30%															
Terminal Value	$ 3,476,250															
Terminal Growth Rate	3%															
NPV	**$2,674,038.46**															
Exit Analysis																
Exit Method	Private Sale															
Exit Multiple	10															
Exit Valuation	**$9,112,500.00**															
Exit Year	2029															
Recurring revenue per user	$ 2,000.00															
Users				100.00	300.00	900.00	1,350.00	2,025.00	3,037.50	4,556.25	4,556.25	4,556.25	4,556.25	4,556.25	4,556.25	4,556.25
Paying users				10.00	30.00	90.00	135.00	202.50	303.75	455.63	455.63	455.63	455.63	455.63	455.63	455.63

DegreeLinked:
Concept: DegreeLinked simplifies the college application process by allowing would be students to complete one digital college application that gets used for all college applications.
Market size: $250 billion (amount spent by universities on acquiring and matriculating students)
Total addressable market: $25 billion (digital advertising to students)
UNIT ECONOMICS (PROFITABILITY) OF THE COMPANY:
a. IDENTIFY COST OF ACQUISITION PER CUSTOMER (CAC): cost to custom build database/# of universities in database = $2,000/14,000 = $.14
b. IDENTIFY COST PER USER PER MONTH: server cost per month/users = $150/14,000 = $.01
c. IDENTIFY REVENUE PER USER PER MONTH: total revenue per year/12 months = $50,000/12 = $4,100
d. IDENTIFY LIFETIME VALUE PER USER (LTV): revenue per user per year x life expectancy = $50,000 x 60 = $3,000,000
MODEL PLATFORM ARCHITECTURE IN TERMS OF BASE PHP PLATFORM AND MODS ONTO IT

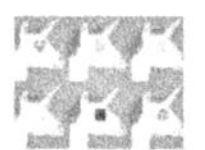

DATABASE OF 25 HIGHEST PRIORITY USERS OF PLATFORM AND COLLECT EMAIL, NAME, NUMBER. .XLS DATABASE NAME:

SELF-ADVOCATE TO THE 25 HIGHEST PRIORITY USERS OF PLATFORM AND NOTIFY THEM THAT YOU WILL BE SENDING THEM VALUE AND THAT THIS IS THEIR ACCOUNT TO CLAIM IT GIVING THEM LOGIN CREDENTIALS. .DOC MESSAGE FILE NAME:

SCREEN SHARE WALK THROUGH WITH THE HIGH PRIORITY USER AND SHOW THEM HOW IT WORKS AND GET THEM TO LOGIN BY THEMSELVES

USER UTILIZES PLATFORM BY THEMSELVES FOR A TRIAL PERIOD OF TIME (3 TO 6 MONTHS)

SEND USER THE INVOICE FOR USING THE PLATFORM. FEE SHOULD BE MEMBERSHIP FEE BASED UPON STANDARD MARKET FEE.

BUILD DATABASE OF 26-50th HIGHEST PRIORITY USERS OF PLATFORM AND COLLECT EMAIL, NAME, NUMBER

IDENTIFY 3 TO 5 MOST LIKELY STRATEGIC BUYERS OF YOUR PLATFORM. REACH OUT TO THE PERSONALLY WHEN YOU HAVE $100M IN RECURRING REVENUE AND EXIT AT THE MARKET REVENUE MULTIPLE BASED UPON PRECEDENT

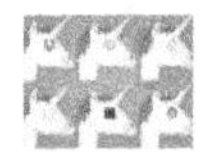

TRANSACTIONS: Facebook, LinkedIn, Google, Alibaba

DISCOUNT RATE STATUS:
Seed: pre-database MVP: 80% discount rate
Angel: self-advocacy & activation: 50%-70% discount rate
Series A: utilization & monetization: 40%-60% discount rate
Series B: monetization & scale: 30%-50% discount rate
Growth: Scale & new markets: 20%-40% discount rate
Mezzanine: Scale in old & new markets: 15%-30% discount rate
IPO: Scale globally: 7-15% discount rate
Post-IPO Minority Ownership: Mature & New Product Development: 3.5%-10% discount rate

FINANCIAL MODEL & VALUATION (DCF):

DegreeLinked

	2014	2015	2016	2017	2018	2019	2020	2021
Membership (recurring revenue)				$ 1,250,000.00	$ 3,750,000.00	$ 11,250,000.00	$ 16,875,000.00	$ 25,312,500.00
Profit per year	**$0.00**	**$0.00**	**$0.00**	**$1,250,000.00**	**$3,750,000.00**	**$11,250,000.00**	**$16,875,000.00**	**$25,312,500.00**
Discount Rate	80%							
Terminal Value	$ 148,296,971							
Terminal Growth Rate	3%							
NPV	**$4,692,341.59**							
Exit Analysis								
Exit Method	IPO							
Exit Multiple	15							
Exit Valuation	**$1,668,548,583.98**							
Exit Year	2029							
Recurring revenue per university	$ 50,000.00							
Universities				25.00	75.00	225.00	337.50	506.25
Paying universities				25.00	75.00	225.00	337.50	506.25

	2022	2023	2024	2025	2026	2027	2028	2029
Membership (recurring revenue)	$ 37,968,750.00	$ 56,953,125.00	$ 71,191,406.25	$ 88,989,257.81	$111,236,572.27	$111,236,572.27	$111,236,572.27	$111,236,572.27
Profit per year	**$37,968,750.00**	**$56,953,125.00**	**$71,191,406.25**	**$88,989,257.81**	**$111,236,572.27**	**$111,236,572.27**	**$111,236,572.27**	**$111,236,572.27**
Universities	759.38	1,139.06	1,423.83	1,779.79	2,224.73	2,224.73	2,224.73	2,224.73
Paying universities	759.38	1,139.06	1,423.83	1,779.79	2,224.73	2,224.73	2,224.73	2,224.73

FundLinked:

Concept: Private equity network and marketplace connecting limited partners (LPs) with general partners (GPs) representing funds being raised.

Market size: $454 billion (total LP commitments)

Total addressable market: $15 bn (fund placement)

UNIT ECONOMICS (PROFITABILITY) OF THE COMPANY:

a. IDENTIFY COST OF ACQUISITION PER CUSTOMER (CAC): database cost/# of companies in database = $3000/12,000 = $.25

b. IDENTIFY COST PER USER PER MONTH: server cost/total users = $150/12,000 = $.02

c. IDENTIFY REVENUE PER USER PER MONTH: revenue per year/12 = $2,000/12 = $166.66

d. IDENTIFY LIFETIME VALUE PER USER (LTV): revenue per year x life expectancy = $2,000 x 60 = $120,000

MODEL PLATFORM ARCHITECTURE IN TERMS OF BASE PHP PLATFORM AND MODS ONTO IT

DATABASE OF 25 HIGHEST PRIORITY USERS OF PLATFORM AND COLLECT EMAIL, NAME, NUMBER. .XLS DATABASE NAME:

SELF-ADVOCATE TO THE 25 HIGHEST PRIORITY USERS OF PLATFORM AND NOTIFY THEM THAT YOU WILL BE SENDING THEM VALUE AND THAT THIS IS THEIR ACCOUNT TO CLAIM IT GIVING THEM LOGIN CREDENTIALS. .DOC MESSAGE FILE NAME:
SCREEN SHARE WALK THROUGH WITH THE HIGH PRIORITY USER AND SHOW THEM HOW IT WORKS AND GET THEM TO LOGIN BY THEMSELVES
USER UTILIZES PLATFORM BY THEMSELVES FOR A TRIAL PERIOD OF TIME (3 TO 6 MONTHS)
SEND USER THE INVOICE FOR USING THE PLATFORM. FEE SHOULD BE MEMBERSHIP FEE BASED UPON STANDARD MARKET FEE.
BUILD DATABASE OF 26-50th HIGHEST PRIORITY USERS OF PLATFORM AND COLLECT EMAIL, NAME, NUMBER
IDENTIFY 3 TO 5 MOST LIKELY STRATEGIC BUYERS OF YOUR PLATFORM. REACH OUT TO THE PERSONALLY WHEN YOU HAVE $100M IN RECURRING REVENUE AND EXIT AT THE MARKET REVENUE MULTIPLE BASED UPON PRECEDENT TRANSACTIONS.
DISCOUNT RATE STATUS: Palico, LinkedIn, Google, Alibaba

Seed: pre-database MVP: 80% discount rate

Angel: self-advocacy & activation: 50%-70% discount rate
Series A: utilization & monetization: 40%-60% discount rate
Series B: monetization & scale: 30%-50% discount rate
Growth: Scale & new markets: 20%-40% discount rate
Mezzanine: Scale in old & new markets: 15%-30% discount rate
IPO: Scale globally: 7-15% discount rate
Post-IPO Minority Ownership: Mature & New Product Development: 3.5%-10% discount rate

FINANCIAL MODEL & VALUATION (DCF):

FundLinked

	2014	2015	2016	2017	2018	2019	2020	2021	2022	2023	2024	2025	2026	2027	2028	2029
Membership (recurring revenue)				$ 20,000.00	$ 60,000.00	$ 180,000.00	$ 270,000.00	$ 405,000.00	$ 607,500.00	$ 911,250.00	$ 911,250.00	$ 911,250.00	$ 911,250.00	$ 911,250.00	$ 911,250.00	$ 911,250.00
Profit per year	**$0.00**	**$0.00**	**$0.00**	**$20,000.00**	**$60,000.00**	**$180,000.00**	**$270,000.00**	**$405,000.00**	**$607,500.00**	**$911,250.00**	**$911,250.00**	**$911,250.00**	**$911,250.00**	**$911,250.00**	**$911,250.00**	**$911,250.00**
Discount Rate	80%															
Terminal Value	$ 1,118,945															
Terminal Growth Rate	3%															
NPV	**$40,918.96**															
Exit Analysis																
Exit Method	IPO															
Exit Multiple	15															
Exit Valuation	**$13,668,750.00**															
Exit Year	2029															
Recurring revenue per user	$ 2,000.00															
Users				100.00	300.00	900.00	1,350.00	2,025.00	3,037.50	4,556.25	4,556.25	4,556.25	4,556.25	4,556.25	4,556.25	4,556.25
Paying users				10.00	30.00	90.00	135.00	202.50	303.75	455.63	455.63	455.63	455.63	455.63	455.63	455.63

Asiansbook:

Concept: "Facebook for Asia"

Market size: $121 billion (Digital advertising)

Total addressable market: $25 bn (digital advertising spend in Asia)

UNIT ECONOMICS (PROFITABILITY) OF THE COMPANY:

a. IDENTIFY COST OF ACQUISITION PER CUSTOMER (CAC): cost of country database/# of consumers in database = $200/1,000,000 = $.02

b. IDENTIFY COST PER USER PER MONTH: $2.40 (Facebook data discounted 80%)

c. IDENTIFY REVENUE PER USER PER MONTH: $3.80 (Facebook data discounted 80%)

d. IDENTIFY LIFETIME VALUE PER USER (LTV): Revenue per user per month x 12 x life expectancy = $3.80 x 12 x 60 = $2,736

MODEL PLATFORM ARCHITECTURE IN TERMS OF BASE PHP PLATFORM AND MODS ONTO IT

DATABASE OF 25 HIGHEST PRIORITY USERS OF PLATFORM AND COLLECT EMAIL, NAME, NUMBER. .XLS DATABASE NAME:

SELF-ADVOCATE TO THE 25 HIGHEST PRIORITY USERS OF PLATFORM AND NOTIFY THEM THAT YOU WILL BE SENDING THEM VALUE AND THAT THIS IS

THEIR ACCOUNT TO CLAIM IT GIVING THEM LOGIN CREDENTIALS. .DOC MESSAGE FILE NAME:
SCREEN SHARE WALK THROUGH WITH THE HIGH PRIORITY USER AND SHOW THEM HOW IT WORKS AND GET THEM TO LOGIN BY THEMSELVES
USER UTILIZES PLATFORM BY THEMSELVES FOR A TRIAL PERIOD OF TIME (3 TO 6 MONTHS)
SEND USER THE INVOICE FOR USING THE PLATFORM. FEE SHOULD BE MEMBERSHIP FEE BASED UPON STANDARD MARKET FEE.
BUILD DATABASE OF 26-50th HIGHEST PRIORITY USERS OF PLATFORM AND COLLECT EMAIL, NAME, NUMBER
IDENTIFY 3 TO 5 MOST LIKELY STRATEGIC BUYERS OF YOUR PLATFORM. REACH OUT TO THE PERSONALLY WHEN YOU HAVE $100M IN RECURRING REVENUE AND EXIT AT THE MARKET REVENUE MULTIPLE BASED UPON PRECEDENT TRANSACTIONS: Facebook, Google, Alibaba, LinkedIn

DISCOUNT RATE STATUS:
Seed: pre-database MVP: 80% discount rate
Angel: self-advocacy & activation: 50%-70% discount rate

Series A: utilization & monetization: 40%-60% discount rate
Series B: monetization & scale: 30%-50% discount rate
Growth: Scale & new markets: 20%-40% discount rate
Mezzanine: Scale in old & new markets: 15%-30% discount rate
IPO: Scale globally: 7-15% discount rate
Post-IPO Minority Ownership: Mature & New Product Development: 3.5%-10% discount rate

FINANCIAL MODEL & VALUATION (DCF):

Asiansbook

	2014	2015	2016	2017	2018	2019	2020	2021	2022	2023	2024	2025	2026	2027	2028	2029
Revenue per user				$ 380.00	$ 3,800.00	$ 38,000.00	$ 380,000.00	$ 3,800,000.00	$ 19,000,000.00	$ 28,500,000.00	$ 42,750,000.00	$ 64,125,000.00	$ 96,187,500.00	$ 144,281,250.00	$ 216,421,875.00	$ 324,632,812.50
Cost per user				$ 240.00	$ 2,400.00	$ 24,000.00	$ 240,000.00	$ 2,400,000.00	$ 12,000,000.00	$ 18,000,000.00	$ 27,000,000.00	$ 40,500,000.00	$ 60,750,000.00	$ 91,125,000.00	$ 136,687,500.00	$ 205,031,250.00
Profit per year	**$0.00**	**$0.00**	**$0.00**	**$140.00**	**$1,400.00**	**$14,000.00**	**$140,000.00**	**$1,400,000.00**	**$7,000,000.00**	**$10,500,000.00**	**$15,750,000.00**	**$23,625,000.00**	**$35,437,500.00**	**$53,156,250.00**	**$79,734,375.00**	**$119,601,562.50**
Discount Rate	80%															
Terminal Value	$ 159,986,506															
Terminal Growth Rate	3%															
NPV	**$4,703,881.74**															
Exit Analysis																
Exit Method	IPO															
Exit Multiple	20															
Exit Valuation	**$2,392,031,250.00**															
Exit Year	2029															
Revenue per user per year	$ 3.80	19														
Cost per user per year	$ 2.40	12														
Users				100.00	1,000.00	10,000.00	100,000.00	1,000,000.00	5,000,000.00	7,500,000.00	11,250,000.00	16,875,000.00	25,312,500.00	37,968,750.00	56,953,125.00	85,429,687.50

FameLinked:

Concept: "Monetizing Social with Microsponsorships" via a ranked social network

Market size: $121 billion (Digital advertising and digital subscriptions for USA businesses)

Total addressable market: $75 billion (digital advertising on social networks)

UNIT ECONOMICS (PROFITABILITY) OF THE COMPANY:

a. IDENTIFY COST OF ACQUISITION PER CUSTOMER (CAC): cost of database/# of consumers in database = $2000/1,000,000 = $.20

b. IDENTIFY COST PER USER PER MONTH: $2.40 (Facebook data discounted 80%)

c. IDENTIFY REVENUE PER USER PER MONTH: $3.80 (Facebook data discounted 80%)

d. IDENTIFY LIFETIME VALUE PER USER (LTV): Revenue per user per month x 12 x life expectancy = $3.80 x 12 x 60 = $2,736

MODEL PLATFORM ARCHITECTURE IN TERMS OF BASE PHP PLATFORM AND MODS ONTO IT

DATABASE OF 25 HIGHEST PRIORITY USERS OF PLATFORM AND COLLECT EMAIL, NAME, NUMBER. .XLS DATABASE NAME:

SELF-ADVOCATE TO THE 25 HIGHEST PRIORITY USERS OF PLATFORM AND

NOTIFY THEM THAT YOU WILL BE SENDING THEM VALUE AND THAT THIS IS THEIR ACCOUNT TO CLAIM IT GIVING THEM LOGIN CREDENTIALS. .DOC MESSAGE FILE NAME:
SCREEN SHARE WALK THROUGH WITH THE HIGH PRIORITY USER AND SHOW THEM HOW IT WORKS AND GET THEM TO LOGIN BY THEMSELVES
USER UTILIZES PLATFORM BY THEMSELVES FOR A TRIAL PERIOD OF TIME (3 TO 6 MONTHS)
SEND USER THE INVOICE FOR USING THE PLATFORM. FEE SHOULD BE MEMBERSHIP FEE BASED UPON STANDARD MARKET FEE.
BUILD DATABASE OF 26-50th HIGHEST PRIORITY USERS OF PLATFORM AND COLLECT EMAIL, NAME, NUMBER
IDENTIFY 3 TO 5 MOST LIKELY STRATEGIC BUYERS OF YOUR PLATFORM. REACH OUT TO THE PERSONALLY WHEN YOU HAVE $100M IN RECURRING REVENUE AND EXIT AT THE MARKET REVENUE MULTIPLE BASED UPON PRECEDENT TRANSACTIONS: Facebook, Google, LinkedIn, Alibaba

DISCOUNT RATE STATUS:
Seed: pre-database MVP: 80% discount rate

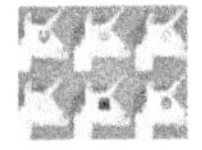

Angel: self-advocacy & activation: 50%-70% discount rate
Series A: utilization & monetization: 40%-60% discount rate
Series B: monetization & scale: 30%-50% discount rate
Growth: Scale & new markets: 20%-40% discount rate
Mezzanine: Scale in old & new markets: 15%-30% discount rate
IPO: Scale globally: 7-15% discount rate
Post-IPO Minority Ownership: Mature & New Product Development: 3.5%-10% discount rate

FINANCIAL MODEL & VALUATION (DCF):

FameLinked

	2014	2015	2016	2017	2018	2019	2020	2021	2022	2023	2024	2025	2026	2027	2028	2029
Revenue per user				$ 900.00	$ 9,000.00	$ 90,000.00	$ 900,000.00	$ 9,000,000.00	$ 45,000,000.00	$ 67,500,000.00	$101,250,000.00	$151,875,000.00	$227,812,500.00	$341,718,750.00	$512,578,125.00	$768,867,187.50
Cost per user				$ 450.00	$ 4,500.00	$ 45,000.00	$ 450,000.00	$ 4,500,000.00	$ 22,500,000.00	$ 33,750,000.00	$ 50,625,000.00	$ 75,937,500.00	$113,906,250.00	$170,859,375.00	$256,289,062.50	$384,433,593.75
Profit per year	$0.00	$0.00	$0.00	$450.00	$4,500.00	$45,000.00	$450,000.00	$4,500,000.00	$22,500,000.00	$33,750,000.00	$50,625,000.00	$75,937,500.00	$113,906,250.00	$170,859,375.00	$256,289,062.50	$384,433,593.75
Discount Rate	80%															
Terminal Value	$ 516,242,340															
Terminal Growth Rate	2%															
NPV	$15,119,019.87															
Exit Analysis																
Exit Method	IPO															
Exit Multiple	20															
Exit Valuation	$7,688,671,875.00															
Exit Year	2029															
Revenue per user per year	$ 9.00															
Cost per user per year	$ 4.50															
Users				100.00	1,000.00	10,000.00	100,000.00	1,000,000.00	5,000,000.00	7,500,000.00	11,250,000.00	16,875,000.00	25,312,500.00	37,968,750.00	56,953,125.00	85,429,687.50

NationLinked:

Concept: It is a government representatives civic duty to represent their constituents and NationLinked is the platform in which constituents directly voice their issues to their representatives. NationLinked is a social governance platform.

Market size: $500 billion (total spend on influencing constituencies)

Total addressable market: $25 billion (total spend on lobbying)

UNIT ECONOMICS (PROFITABILITY) OF THE COMPANY:

a. IDENTIFY COST OF ACQUISITION PER CUSTOMER (CAC): cost to build federal/state/local governance database/# of entries in database = $2,500/2000 = $.25

b. IDENTIFY COST PER USER PER MONTH: $2.40 (Facebook data discounted 80%)

c. IDENTIFY REVENUE PER USER PER MONTH: $3.80 (Facebook data discounted 80%)

d. IDENTIFY LIFETIME VALUE PER USER (LTV): Revenue per user per month x 12 x life expectancy = $3.80 x 12 x 60 = $2,736

MODEL PLATFORM ARCHITECTURE IN TERMS OF BASE PHP PLATFORM AND MODS ONTO IT

DATABASE OF 25 HIGHEST PRIORITY USERS OF PLATFORM AND COLLECT

EMAIL, NAME, NUMBER. .XLS DATABASE NAME:
SELF-ADVOCATE TO THE 25 HIGHEST PRIORITY USERS OF PLATFORM AND NOTIFY THEM THAT YOU WILL BE SENDING THEM VALUE AND THAT THIS IS THEIR ACCOUNT TO CLAIM IT GIVING THEM LOGIN CREDENTIALS. .DOC MESSAGE FILE NAME:
SCREEN SHARE WALK THROUGH WITH THE HIGH PRIORITY USER AND SHOW THEM HOW IT WORKS AND GET THEM TO LOGIN BY THEMSELVES
USER UTILIZES PLATFORM BY THEMSELVES FOR A TRIAL PERIOD OF TIME (3 TO 6 MONTHS)
SEND USER THE INVOICE FOR USING THE PLATFORM. FEE SHOULD BE MEMBERSHIP FEE BASED UPON STANDARD MARKET FEE.
BUILD DATABASE OF 26-50th HIGHEST PRIORITY USERS OF PLATFORM AND COLLECT EMAIL, NAME, NUMBER
IDENTIFY 3 TO 5 MOST LIKELY STRATEGIC BUYERS OF YOUR PLATFORM. REACH OUT TO THE PERSONALLY WHEN YOU HAVE $100M IN RECURRING REVENUE AND EXIT AT THE MARKET REVENUE MULTIPLE BASED UPON PRECEDENT TRANSACTIONS: Facebook, LinkedIn, Google, Alibaba

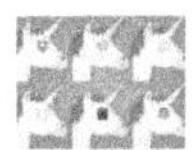

DISCOUNT RATE STATUS:
Seed: pre-database MVP: 80% discount rate
Angel: self-advocacy & activation: 50%-70% discount rate
Series A: utilization & monetization: 40%-60% discount rate
Series B: monetization & scale: 30%-50% discount rate
Growth: Scale & new markets: 20%-40% discount rate
Mezzanine: Scale in old & new markets: 15%-30% discount rate
IPO: Scale globally: 7-15% discount rate
Post-IPO Minority Ownership: Mature & New Product Development: 3.5%-10% discount rate

FINANCIAL MODEL & VALUATION (DCF):

Nationlicked

	2014	2015	2016	2017	2018	2019	2020	2021	2022	2023	2024	2025	2026	2027	2028	2029
Revenue per user				$ 900.00	$ 9,000.00	$ 90,000.00	$ 900,000.00	$ 9,000,000.00	$ 13,500,000.00	$ 20,250,000.00	$ 30,375,000.00	$ 45,562,500.00	$ 68,343,750.00	$102,515,625.00	$153,773,437.50	$230,660,156.25
Cost per user				$ 450.00	$ 4,500.00	$ 45,000.00	$ 450,000.00	$ 4,500,000.00	$ 6,750,000.00	$ 10,125,000.00	$ 15,187,500.00	$ 22,781,250.00	$ 34,171,875.00	$ 51,257,812.50	$ 76,886,718.75	$115,330,078.13
Profit per year	**$0.00**	**$0.00**	**$0.00**	**$450.00**	**$4,500.00**	**$45,000.00**	**$450,000.00**	**$4,500,000.00**	**$6,750,000.00**	**$10,125,000.00**	**$15,187,500.00**	**$22,781,250.00**	**$34,171,875.00**	**$51,257,812.50**	**$76,886,718.75**	**$115,330,078.13**
Users				100.00	1,000.00	10,000.00	100,000.00	1,000,000.00	1,500,000.00	2,250,000.00	3,375,000.00	5,062,500.00	7,593,750.00	11,390,625.00	17,085,937.50	25,628,906.25

Discount Rate	50%
Terminal Value	$ 354,272,702
Terminal Growth Rate	3%
NPV	**$4,536,082.67**

Exit Analysis	
Exit Method	IPO
Exit Multiple	20
Exit Valuation	**$2,306,601,562.50**
Exit Year	2029

Revenue per user per year	$ 9.00
Cost per user per year	$ 4.50

www.ingramcontent.com/pod-product-compliance
Lightning Source LLC
Chambersburg PA
CBHW072038190526
45165CB00018B/1076

9781502408075